Dedication:

For the Lion Recovery Fund and the team at Wildlife Conservation Network, whose tireless work helps wild lions roam free, thrive, and roar for generations to come.
–Jenny Curtis

For Bella and Zoey; one with the pride of a lion, and one with the alertness.
–Allyson Randa

Author: Jenny Curtis

Designer: Allyson Randa

Photo Credits:

AdobeStock.com

Pixabay.com

Pexels.com

ISBN: 978-1-965081-18-1

This book meets Common Core and Next Generation Science Standards.

TABLE OF CONTENTS

THE KING OF THE BEASTS

Step into the grasslands of Africa, and you might spot a tawny figure with a flowing mane or a sleek golden body stalking through the tall grass. That's a lion—one of the most powerful animals on Earth!

DID YOU KNOW?

Lions are known for their strength, teamwork, and a mighty roar you can hear from five miles away.

They live in groups called prides, where everyone has a role—from hunting to guarding to taking care of cubs.

RIB-RATTLING ROAR

Imagine a sound so loud it rattles your ribs. That's a lion's roar! It's one of the loudest sounds in the animal kingdom—louder than a lawn mower or a jackhammer. A roar can reach up to 114 decibels and travel as far as 5 miles away!

The lion's body helps make this mighty noise. It has a special bone in its throat called a hyoid bone, which is flexible and lets the lion make deep, growling roars that other cats can't.

Why do lions roar? Lots of reasons. They roar to say, "This is my land!" or to find other members of their pride. Sometimes, a lion's roar even warns rivals to stay away.

FUN FACT:
Only lions, tigers, jaguars, and leopards can roar!

Their ears can swivel to find the exact direction of sound.

A lion's mane shows strength and protects their neck during fights.

Big eyes help them see well in the dark when they hunt.

DID YOU KNOW?
Lion claws can grow up to 1.5 inches long—about the length of a house key!

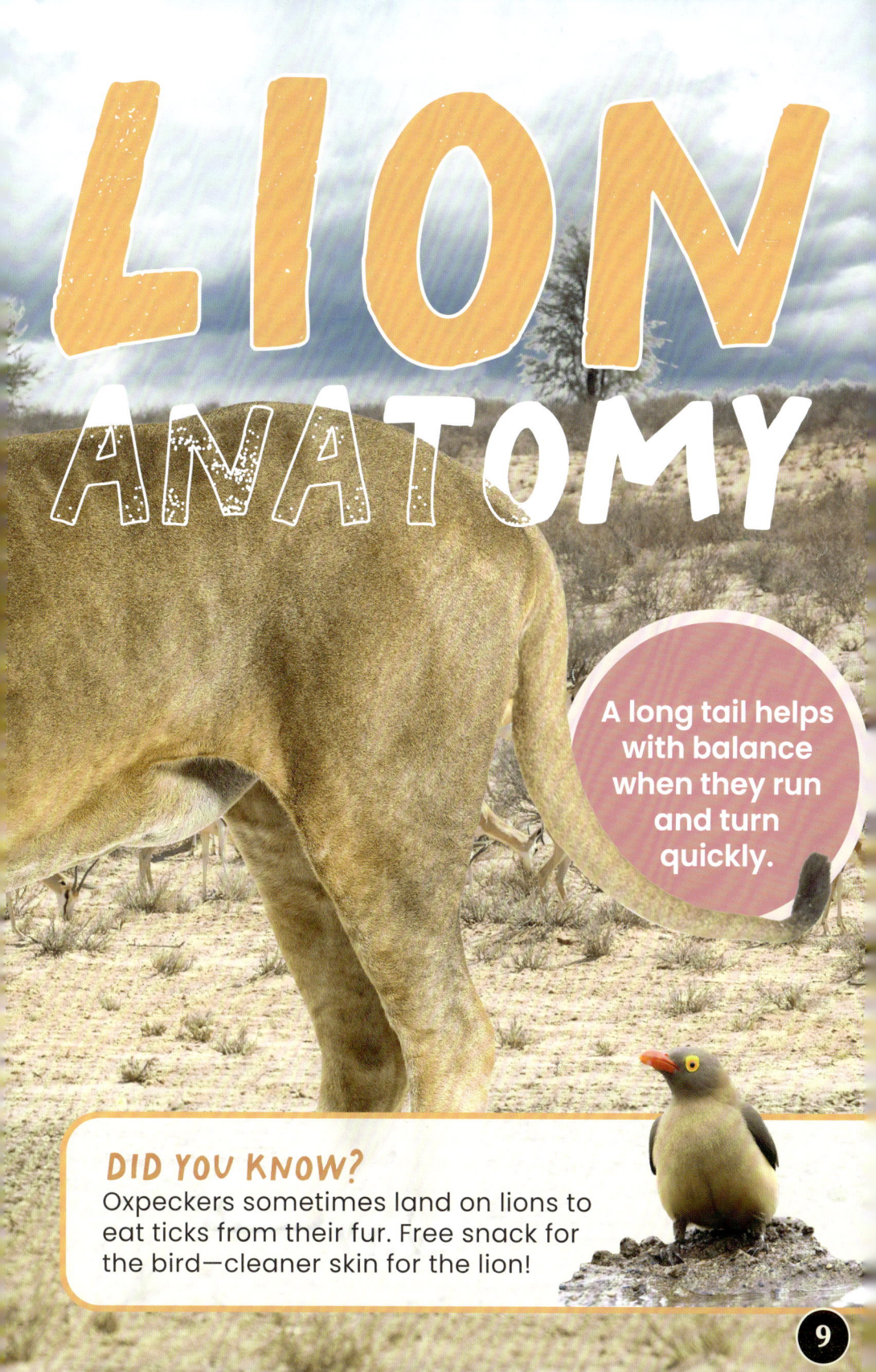

LION ANATOMY

A long tail helps with balance when they run and turn quickly.

DID YOU KNOW?

Oxpeckers sometimes land on lions to eat ticks from their fur. Free snack for the bird—cleaner skin for the lion!

MARVELOUS

Only male lions have manes—and wow, what a mane it is! Thick and fluffy, it surrounds the lion's head. Scientists believe the mane tells other lions how strong and healthy a male is.

Manes also protect lions during battles. When lions fight, they often go for the head and neck. The thick mane acts like a built-in shield! Cubs are born without manes, but they start growing around one year of age.

MANE

Not all manes are alike. Some male lions in very hot places grow only tiny manes—or none at all! That helps them stay cool under the sun.

LION EYES

Lions are excellent night hunters, and their eyes are a big reason why. A special layer behind their eyes—called the tapetum lucidum—reflects light like a mirror. It helps them see clearly, even in near darkness!

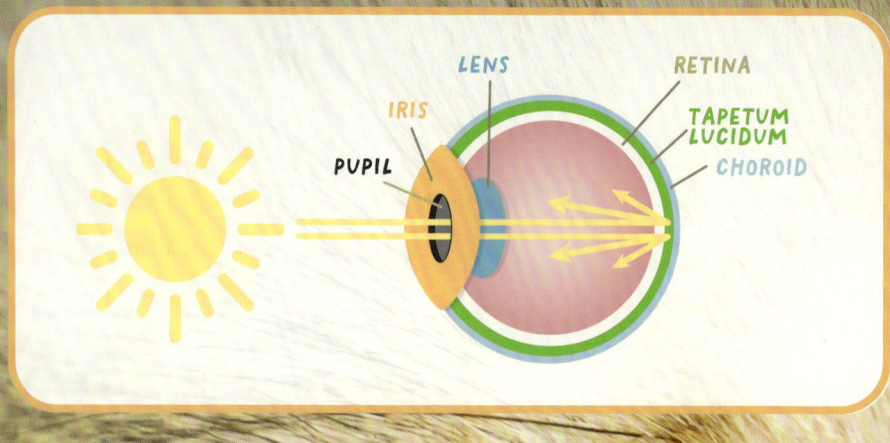

If you've ever seen a lion at night, you might have noticed its glowing eyes. That glow is the tapetum in action, reflecting moonlight back out!

MIGHTY PAWS

Lions are big, but you'd be surprised how quietly they move. That's because their paws are covered in soft, leathery pads that muffle sound. They can sneak through tall grass without being heard by their prey.

Their large paw size also helps them move across rough terrain. And like house cats, they use their paws to clean their faces—yes, even lions like a little grooming time!

FELINE FAMILY TREE

Lions belong to a big cat family called Felidae. That means they're cousins with cheetahs, leopards, and even the tiny wildcats that look like house cats.

CHEETAH

CHEETAHS – Fastest land animals on Earth! They hunt alone, while lions hunt in teams.

LEOPARD

LEOPARDS – Super sneaky and great climbers. They often avoid lions by hiding in trees.

FUN FACT: Leopards drag their food up into trees so lions don't steal it.

CARACAL

CARACALS _and_ SERVALS –
Medium-sized cats with big
ears and great jumping skills!

FUN FACT: Caracals have long
black tufts on their ears that
help them communicate.

SERVAL

AFRICAN WILDCATS

AFRICAN WILDCATS –
Super stealthy and mostly
come out at night.

JAGUAR

JAGUARS –
The biggest cats in the
Americas and the third
largest in the world.

FUN FACT: Jaguars
have the strongest
bite of all big cats!

LION TAXONOMY

Kingdom: Animalia
(*An-i-MALE-yuh*) All animals on Earth are in this kingdom.
Animalia: Living creature.

Phylum: Chordata
(*Kor-daa-tuh*) Animals with a spinal cord are grouped in this phylum.
Chordata: Having a spinal cord.

Class: Mammalia
(*Ma-MALE-yuh*) This level includes warm-blooded animals with hair or fur that nurse their young.
Mammalia: Mamma.

Order: *Carnivora*
(*Car-NIV-or-uh*) Meat-eaters with strong teeth and claws.
Carnivora: Meat-eaters.

Family: Felidae
(*FEE-lih-day*) All cat species, big and small.
Felidae: Cats.

Genus: *Panthera*
(*Pan-THER-uh*) The big cats that can roar! Includes lions, tigers, leopards, and jaguars.
Panthera: Roaring cats.

Species: *Panthera leo*
(*Pan-THER-uh LAY-oh*) This is the lion's full scientific name.
Panthera leo: Lion.

"WHO'S WHO" IN THE LION WORLD

Most people think all lions look the same—but did you know there are several types of lions? Scientists call these subspecies. All of them are part of the *Panthera leo* species, but they've adapted to different parts of the world.

ASIATIC LION

Found only in India's Gir Forest.

Smaller manes and less social than African lions.

WEST AFRICAN LION

 Lives in countries like Senegal and Nigeria.

Very rare, with only a few hundred left in the wild.

 Spotted in Kenya and Tanzania.

Likely what you'll see on a safari in the Serengeti.

EAST AFRICAN LION

LIVING IN A PRIDE

What makes prides special? Most big cats live alone, but lions live, hunt, eat, and sleep together. The females are usually sisters, cousins, or moms and daughters—and they stick together for life!

Pride life is full of teamwork. Female lions hunt together, often using group tactics to surround their prey. After the hunt, everyone eats—but there's a pecking order!

And even though the males don't hunt much, they're important too. Their roars scare away intruders, and their size keeps enemies in check.

LITTLE LION CUBS

When lion cubs are born, they're tiny—only about 3 pounds! Their eyes are closed for the first few days, and they can't walk very well yet. Lion moms (called lionesses) keep their babies hidden in tall grass or bushes to protect them from danger.

SPOTTY CUBS

Lion cubs are born with spots! These help them blend in while hiding.

Cubs don't just play for fun—it's serious practice! Wrestling, pouncing, and chasing are how they learn the skills they'll need to survive as adult lions.

They play with their brothers, sisters, and other cubs in the pride. Older cubs even help care for the younger ones, like big siblings on babysitting duty.

DID YOU KNOW?

Cubs can't roar until they're about a year old—they sound more like squeaky kittens!

WHAT'S FOR

Lions are carnivores—they eat meat, and lots of it! But they don't eat every day. A big lion feast might last a few days, followed by a nice long nap.

ZEBRAS

Zebras – fast and tasty

Wildebeest – a top choice during migrations

WILDEBEEST

BUFFALO

Buffalo – big, strong, and risky but rewarding

Gazelles – small and speedy

GAZELLES

DINNER?

ANTELOPE

Antelope – common targets

Warthogs – quick but meaty

WARTHOGS

ANYTHING ELSE?

Birds – a snack if they get close

Reptiles – rarely, but possible

Monkeys – if they can catch them

GROUP DINNER

Lions don't use forks and knives—but they've got something better: their teeth, tongues, and teamwork! A lion's jaws are strong enough to crush bone, and their teeth are shaped to cut meat like scissors.

DIET DATA

90

Lions need 11–15 pounds of meat per day (more for males)

A full-grown lion can eat up to 90 pounds in one meal!

They may go 3–4 days without eating after a big feast

THE FO

Antelope

Lions eat meat and help control animal populations.

OD WEB

Vultures eat leftovers and help keep the land clean.

Zebras and antelope eat grass and help spread seeds.

Use this as a guide for your ecosystem poster

LIONS ON PATROL

Lions don't just sleep all day—some walk miles to protect their territory!

Male lions patrol the edges of their land to keep their pride safe. They sniff trees, rub their faces on bushes, and even pee on tall grass to leave scent marks that say, "This place is taken!

FUN FACT

Lions have scent glands in their paws and faces—they can mark territory just by walking or rubbing on a tree!

Cubs are born with blue eyes, which change to golden-brown as they grow.

Lions can leap the length of a school bus in one jump!

Lions are the only big cats that live in social groups!

Manes make lions look bigger to enemies—like wearing a puffed-up coat of armor.

DID YOU

Lions and hyenas often steal each other's food—it's a wild rivalry!

Grooming each other is called allogrooming—and lions love it!

Lions share 95.6% of their DNA with house cats!

Lions have five toes on their front paws, but only four on the back.

KNOW?

You'll find lions snoozing in the shade, draped across each other like cuddly pillows. Cubs nap between bursts of play, and adults stretch out in the dirt or under a tree.

But even while resting, lions stay alert. One ear may twitch, or a tail may flick if something interesting happens nearby. If danger comes? They're up in a flash!

MEET
THE LION RECOVERY FUND

The African savanna is alive with the deep roar of lions and the rhythms of local villages. Lions and people share this landscape, their futures are intertwined. But across 52% of Africa's lion range, populations are declining due to poaching, habitat loss, and conflict. The Lion Recovery Fund (LRF) focuses on three pillars—lions, landscapes, and livelihoods—working to balance conservation with the needs of local communities.

Photos from: Lion Recovery Fund

Thanks to your support, the LRF funds diverse initiatives turn conflict into coexistence - from restoring habitats to creating new livelihoods. Across Kenya, Mozambique, South Sudan, and beyond, these efforts are showing us how protecting lions also preserves ecosystems and empowers communities.

HOW SCIENTISTS TRACK LIONS

How do scientists learn what lions do all day? They use high-tech tracking collars!

These collars have tiny GPS devices inside. When a lion wears one, it sends signals to a computer that show where the lion goes—day or night. This helps scientists learn where lions eat, sleep, and roam.

FUN FACT

A GPS collar can track a lion's location every 30 minutes—some even have solar panels to keep them powered by sunlight!

Tracking lions is a smart way to protect them. If a lion gets too close to people or livestock, rangers can step in before there's trouble. That keeps both lions and communities safe.